DESIGN AND ENGINEERING

GADGETS AND GAMES

Chris Oxlade

Heinemann
LIBRARY

Chicago, Illinois

Edited by Andrew Farrow, Abby Colich, and
 Vaarunika Dharmapala
Designed by Richard Parker
Original illustrations © Capstone Global Library
 Ltd 2013
Illustrations by HL Studios
Picture research by Elizabeth Alexander
Originated by Capstone Global Library Ltd
Printed and bound in China by CTPS

16 15 14 13 12
10 9 8 7 6 5 4 3 2 1

Library of Congress Cataloging-in-Publication Data
Oxlade, Chris.
 Gadgets and games / Chris Oxlade.
  p. cm.—(Design and engineering for STEM)
 Includes bibliographical references and index.
 ISBN 978-1-4329-7031-4 (hb)—ISBN 978-1-4329-7036-9
(pb) 1. Toys—Design and construction—Juvenile literature.
2. Household appliances—Design and construction—Juvenile
literature. 3. Electronic apparatus and appliances—Design
and construction—Juvenile literature. 4. Product life cycle—
Juvenile literature. I. Title.

TS2301.T7O95 2013
688.7'2—dc23   2012013468

Acknowledgments
We would like to thank the following for permission
to reproduce photographs: Alamy pp. 5 top right (©
INTERFOTO), 12 (© LOOK Die Bildagentur der Fotografen
GmbH), 15 (© Hugh Threlfall), 22 (© Picture Contact BV), 24
(© David Stock), 31 (© Ryan McGinnis), 36 (© Corbis Bridge),
37 (© James Blinn); Capstone Global Library/ Lord and
Leverett p. 29; Corbis pp. 26 (© Oliver Berg/DPA), 32 (© Qilai
Shen/In Pictures), 43 (© Sean Yong/Reuters); Getty Images pp.
11 (© Jay P. Morgan/Brand X Pictures), 30 (K. Beebe/Custom
Medical Stock Photo), 35 (Leo Ramirez/AFP), 38 (PARK
JI-HWAN/AFP), 40 (Oli Scarff), 46 (Issouf Sanogo/AFP);
Image courtesy of iFixit, www.ifixit.com p. 19; Images courtesy
of Daniel Boardman, BMS Design Ltd, www.bmsdesignltd.
co.uk p. 14; Reproduced with permission of ANYS, Inc. p.
18; Science Photo Library pp. 17 (Simon Fraser/Welwyn
Electronics); 48 (Jerry Mason); Shutterstock pp. 5 bottom
right (© scyther5), 5 left (© Andrew Buckin), 8 (© cobalt88),
8 (© Spectrum7), 8 (© Seregam), 9 (© Sabri Deniz Kizil), 9 (©
tele52), 9 (© tele52), 11 & 47 (© cobalt88), 20 (© NicoTucol),
23 (© Alex Mit), 25 top (© Leenvdb), 25 bottom (©
koya979), 39 (© rangizzz), 42 (©Blazej Lyjak), 45 top (© Olga
Drabovich), 45 bottom (© Makhnach), 49 (© Steve Mann);
design feature arrows Shutterstock (© MisterElements).

Cover photograph of a man hand holding a cell phone
reproduced with permission of Superstock (© Christin
Gilbert/Age Fotostock).

Every effort has been made to contact copyright holders
of material reproduced in this book. Any omissions will
be rectified in subsequent printings if notice is given to
the publisher.

All the Internet addresses (URLs) given in this book were
valid at the time of going to press. However, due to the
dynamic nature of the Internet, some addresses may have
changed, or sites may have changed or ceased to exist since
publication. While the author and publisher regret any
inconvenience this may cause readers, no responsibility
for any such changes can be accepted by either the author or
the publisher.

CONTENTS

Some words are shown in bold, **like this**. You can find out what they mean by looking in the glossary.

GADGETS, GAMES, AND LIFE CYCLES

Gadgets are part of our everyday lives. The most popular gadgets include cell phones, **smartphones** (such as the iPhone and Blackberry phones), MP3 players, tablets (for example, the iPad), and e-book readers (such as the Kindle). In this book, we also look at video game consoles, such as the Xbox and PlayStation. Have you ever wondered how these gadgets are made?

All gadgets are made up of hardware (their physical parts) and **software**. A gadget needs software called an **operating system** (such as iOS) that controls the parts of the gadget itself. It also needs programs or **applications** ("apps" for short), which make it do different jobs, such as e-mailing and browsing the Internet. Games for computers, video game consoles, tablets, and smartphones are applications.

Gadget life cycles

All gadgets have life cycles. The cycle begins with the initial idea for the gadget and ends when it is no longer used. In between these come **designing**, making **prototypes**, **manufacturing**, **marketing** and selling, maintaining, and finally reusing, **recycling**, or disposing. The life cycle of some gadgets, such as smartphones, is very short—it might be just a couple of years. This is because many people replace their phones every year or so.

LIFE CYCLE PEOPLE

Many people are involved in the life cycle of a gadget. They include designers and engineers, as well as the people who assemble and sell the product. Designers and engineers work on the design, prototyping, and manufacturing parts of a **product life cycle**, which is called the **engineering** life cycle.

Designers and engineers often invent new **technologies**, which are then used in gadgets. They also improve technology that already exists, to make gadgets better or cheaper. You can find out more about careers in engineering and design on pages 54-55.

Gadgets old and new

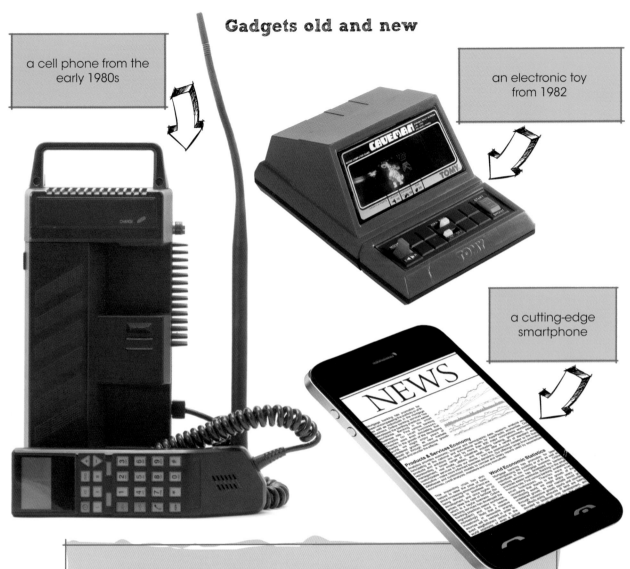

a cell phone from the early 1980s

an electronic toy from 1982

a cutting-edge smartphone

CONTRAST THE PAST

The smartphones, video game consoles, MP3 players, and tablets we have today are very different from the first electronic gadgets. This is evidence for how quickly technology has changed. The first hand-held cell phones appeared in the 1970s. They were the size of a brick, and almost as heavy! They also only made phone calls.

Modern smartphones are really compact computers, with the power that desktop personal computers had just a decade ago. Miniaturization (see page 17) of electronics has been one factor in this change. The first computer game consoles also appeared in the 1970s. They allowed you to play very simple, two-dimensional (2-D) tennis games. Modern consoles have enormous power by comparison, with the ability to display incredibly detailed three-dimensional (3-D) graphics.

Important terms

There are many terms and concepts that are used in the world of design, engineering, and technology. They are also referred to as we look at the life cycle of gadgets and games. Knowing these terms and concepts will help you to understand all product life cycles, not just those of gadgets and games. The terms are summarized here so you can refer back to them when you need to.

Product life cycle

A product life cycle refers to the series of events in the life of a gadget. It starts with the initial idea and ends with disposal or recycling. The life cycle includes design, prototyping, manufacturing, and marketing.

Design

Design involves the process of choosing materials, **components**, and the appearance of a gadget, taking into account **requirements** and **constraints** in a design brief. **Software design** is figuring out how software for gadgets and games will work.

Requirements

Requirements are the features and characteristics a gadget must have (such as speed, memory size, or battery life). Requirements are written in a design brief.

Constraints

Constraints are the limits the designers have to work within (such as budget and size). Constraints are written in a design brief.

System

A **system** is a set of things that work together to do a job. A system has an **input**, a **process**, and an **output**. Information of some sort goes into the system through the input, is processed, and then goes to an output.

Prototype

A prototype is a test version, made before the gadget goes into production, to make sure it can be made and works properly.

Engineering

Engineering is using science and technology to design and make machines, structures, and devices. Electronic software and production engineering is particularly important in the design and manufacturing of gadgets.

Manufacturing

Manufacturing is the process of making products from materials and components. Gadgets are normally made by **mass production**, on **assembly lines**. Software DVDs and Blu-Ray discs are made by batch production, where limited quantities of identical objects are made at the same time.

Computer-aided design and manufacturing

Computer-aided design (CAD) and **computer-aided manufacturing (CAM)** are the use of information and computer technology in the design and manufacturing of products.

Software

Software includes instructions and data that control what a gadget does. System software controls the parts of the gadget itself. Application software (an app) makes the gadget do a specific job.

Marketing

Marketing involves bringing a gadget to the attention of the consumer by advertising and holding product launches. Marketing also includes market research, which tries to find out what gadgets people want to buy.

Recycling

Recycling means using materials again. When a gadget reaches the end of its life cycle, it is recycled to reuse materials such as plastics and metals.

WHAT IS TECHNOLOGY?

Technology is modifying objects and materials to satisfy people's needs and wants. Technology can be as simple as a wooden stool and as complex as an **integrated circuit**.

THE LIFE CYCLE OF A TABLET

This flow chart shows the stages in the life cycle of a typical tablet, as an example of a gadget life cycle.

Initial idea

The first step in a tablet's life cycle is the initial idea, which comes from the manufacturer. The idea may come in response to consumer demand, or because the manufacturer has identified a new market. A design brief describes the requirements and constraints of the gadget.

Design

Working from a design brief, the designers choose materials and components and design the appearance, functions, and software. They write a design **specification** for the complete gadget.

End of life

Eventually, all gadgets become unused, outdated, or broken beyond repair. One of the following happens:

- Reusing: The tablet is given to another user, often after being refurbished.
- Recycling: The tablet is broken up and the materials inside are recycled. Some materials may be used as raw materials for new gadgets.
- Disposal: The tablet is thrown away.

Prototyping

A prototype is made to make sure the design works and that all the components fit and work together. Changes might be made to the design specification at this stage to solve any problems.

Materials

Extracting materials from the ground and processing them is part of the tablet's life cycle (for example, mining aluminum ore and processing it to extract the aluminum).

Manufacturing

Mass production is carried out on an assembly line. Components are added to the gadget until it is complete. Assembly starts with the circuit board. Electronic components are added to the board, then it is fitted into the case with other parts, such as the screen and connectors. The tablets are packaged, ready for sale.

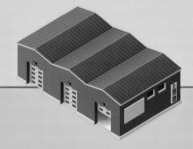

Marketing

The manufacturer advertises on the Internet, on television, and in newspapers. The manufacturer holds a product launch party. Finished gadgets are distributed to retail stores and online sellers.

Useful life

Consumers buy the tablet. This stage is the **useful life** of the tablet. The tablet must be maintained throughout this period. Maintenance includes customer support, making spare parts, repairing flawed tablets, and releasing software updates.

WHAT IS THE NEED?

The life cycle of gadgets and games begins with an idea for a new product. Companies that manufacture these products want to make money. They will only spend money and time making them when they think there is a market for them. This means people want them and will probably buy them.

Market pull ...

Consumers want gadgets because they are useful for communicating, for entertainment, for navigating, for reading e-books, and for work. Consumers want new games to play for entertainment and fun. This need from people is called consumer demand, or "market pull." Many people want the latest gadgets. These people create more demand, because they regularly buy new gadgets. Video game fans also want the latest games to play.

 ... and market push

Manufacturers also drive the need for new gadgets and games by manufacturing products they think consumers will want to buy. This is known as "market push." They might design a gadget with new features or that makes use of a new technology (such as a 3-D screen), or they might improve an existing gadget. They may also introduce new products in response to their rivals or update an existing product to extend its life cycle until a major new product is launched. Manufacturers always think about whether there is a gadget that does not exist that people will want—this is known as spotting a hole in the market.

MARKET RESEARCH

Manufacturers talk to people about their products. They ask them questions about which features they like, which ones they do not, and what they would like to use in the future. This is known as market research. Market researchers also look at sales figures to see which products are selling well and to look for holes in the market. Research helps manufacturers understand the latest trends in technology and how these might be relevant to them.

Uses of cell phones

text messaging	**43%**
safety	**35%**
to keep in touch with friends	**34%**
to keep in touch with family	**26%**
to always be in contact	**22%**
for convenience	**20%**
my friends have one	**17%**
so I don't have to borrow one	**15%**
so I don't have to use the family home phone	**11%**
privacy	**10%**

This bar chart shows reasons for buying a cell phone as given by American teenagers in a 2010 poll.

One very important function of a cell phone is providing safety. It allows people to let their friends and family know where they are at all times.

New technologies

Inventors and engineers are developing new technologies for gadgets all the time. Engineers also improve technologies that already exist and find ways to manufacture technologies more cheaply. New technologies allow new or improved gadgets to be designed and manufactured. This effect is called "technology push." Examples include the increasing speed of mobile data networks (such as the introduction of **4G** networks), which allow smartphones and tablets to communicate faster, and **cloud computing**. Many new technologies are developed by the research and development departments of major companies.

Since the success of the Nintendo Wii, other manufacturers have developed their own motion control systems.

THE NINTENDO WII

The Nintendo Wii was launched at the end of 2006. By the end of 2007, consumers around the world had bought 20 million units. By March 2011, 86 million Nintendo Wii units had been sold worldwide. The Wii's success was due mainly to its motion controllers. These contain sensors that send information about the position and angle of the controller to the console and allow players to control games with hand and arm movements.

Touch-screen displays, 3-D screens, motion sensors, lithium-ion batteries, digital compasses, face recognition, speech recognition, and wireless networking are all technologies that were once new, but are now common in modern gadgets. All these technologies make gadgets more useful and easier to use. New types of software developed for video games are also classed as new technology.

Technology push may also come from the need to reduce harm to the environment or to meet certain standards set by a country's laws. For example, solar-powered rechargers for gadgets and gadget batteries are a technology that reduces electricity consumption.

The new technologies used in gadgets are expensive to produce at first, so the gadgets that use them are expensive to buy. Prices gradually fall as the gadgets are made and sold in greater quantities. For example, the first touch-screen smartphones were expensive, but now even low-budget smartphones have touch screens.

LIFE CYCLE LENGTHS

The new technology in gadgets is developed very quickly. For example, more powerful processors are introduced every few months. This means some gadgets—especially cell phones—become out-of-date very quickly, often in just a few months. So, the lifecycle of these gadgets is shorter than other products, such as televisions and cars.

WHAT HAVE WE LEARNED?

- The life cycle of a gadget begins with a need.
- Market pull is demand from consumers.
- Market push is created by manufacturers.
- Market research finds out what consumers want.
- New technologies allow new types of gadgets to be developed.

GADGET DESIGN

Once a gadget manufacturer has identified the need for a new gadget, the next step is to design it. The design stage is when an idea begins to be turned into a real product. Gadget designers are given a design brief that lists the requirements for a product and also the constraints.

This is a model for a new smartphone design.

REQUIREMENTS AND CONSTRAINTS

Requirements for a smartphone might be:

- the materials for the case
- screen size
- battery life
- resolution of the camera.

Constraints might be:

- the budget (how much the phone will cost to manufacture, which will limit the materials and components that the designers can choose)
- the maximum case size.

Manufacturers have to design a gadget's electronics, its software, and its shape, look, feel, and buttons. They employ designers from different disciplines (such as electronics designers, software designers, and artistic designers). The designers work from the design brief. They use their knowledge and experience of designing products in the past to design the new gadget.

Materials and components

Designers choose the materials a gadget will be made from. These will include plastics and metals for the structure, case, and buttons. Materials are chosen for their properties, such as strength, durability, color, and texture. For example, many handheld gadgets have a case made from polycarbonate, which is a very tough plastic, in case they are dropped. Designers also choose the components for gadgets, such as buttons, touch screens, and electronic components such as microprocessors and memory chips. The chosen materials and components form part of the design specification for a product. This contains all the information needed to manufacture the product.

Jonathan Ive is head designer at Apple. He designed the iPod, iPhone, and iPad, which are regarded as design classics. He says: "Fanatical attention to detail and coming across a problem and being determined to solve it is critically important."

ECO IMPACT

Gadgets use up energy throughout every stage in their life cycle. Much of it comes from burning fuels, which releases carbon dioxide and other greenhouse gases into the atmosphere. The amount of these gases released in the life cycle of a product is called its carbon footprint.

Recycled materials and energy efficiency reduce the carbon footprint of a gadget. For example, a Nokia 700 smartphone contains about 30 percent recycled materials and uses energy-saving technologies. It has a carbon footprint of 20 pounds (9 kilograms) of carbon dioxide. That is the same as the carbon footprint of 60 cans of soda.

Gadget systems

A system is a set of things that work together to do a job. A system has an input, a process, and an output. Information goes into the system through the input. It goes through a process, and it then goes to an output.

Gadgets have complex systems that continuously take inputs from buttons, touch screens, and microphones, process this information, and continuously send it to the screen, speakers, and other outputs. The system is controlled by software. All the parts of a gadget's system have to be designed. Sometimes designers take apart products from other manufacturers to figure out their systems. This is known as reverse engineering.

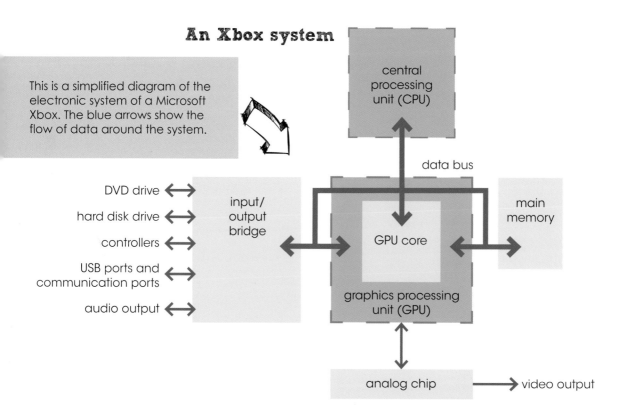

An Xbox system

This is a simplified diagram of the electronic system of a Microsoft Xbox. The blue arrows show the flow of data around the system.

central processing unit (CPU)

data bus

DVD drive
hard disk drive
controllers
USB ports and communication ports
audio output

input/output bridge

GPU core

graphics processing unit (GPU)

main memory

analog chip → video output

INTERFACE DESIGN

Part of designing a gadget's system is the interface between it and the user. For example, an e-book reader's system allows the user to flip from one page of a book to the next. The system of touches and swipes on the e-book reader's screen has to be designed, as does the way the pages of the book are displayed. Good interface design makes a gadget easy to operate.

Electronic design

Electronics designers design the complex electronic circuits inside gadgets. These are made up of circuit boards with components attached to them. The components include:

- buttons
- microphones
- speakers
- integrated circuits such as memory chips, microprocessors, and graphics controllers.

All these circuits are powered by a battery. Some new gadgets will use circuits from existing gadgets, and some will have circuits that are designed from scratch.

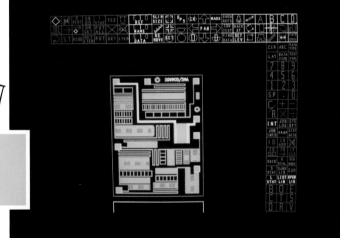

This integrated circuit is being designed on a computer.

MICROSCOPIC COMPONENTS

Every new generation of gadgets is faster and has more features than the previous generation. The power of software processors and the capacity and speed of memory chips increases from year to year. Designers are also able to squeeze the electronics into smaller spaces. This is because of the increasing miniaturization of components on integrated circuits. In modern integrated circuits, more than a million components can be squeezed into a single square millimeter.

There is a rule of thumb called "Moore's Law," thought up by the cofounder of the Intel company, Gordon E. Moore, that says the number of transistors placed inexpensively on an integrated circuit doubles approximately every two years. This has been adapted to say that the chip performance doubles approximately every 18 months.

Computers in design

Information and communications technology is an important tool. Working from a design specification, designers can draw the parts of gadgets on a computer. This is known as computer-aided design (CAD). Designers draw what the parts look like from the sides, from the ends, and from the top and bottom. The computer software builds up 3-D models that can be viewed on a screen. The parts can be displayed in "wire-frame," which means its shape is drawn with many polygons.

Models can also be displayed looking solid, in color, lit up from different directions, and with texture added. They can be tilted and turned and viewed from any angle. Realistic models are particularly useful for designing the appearance of a new gadget.

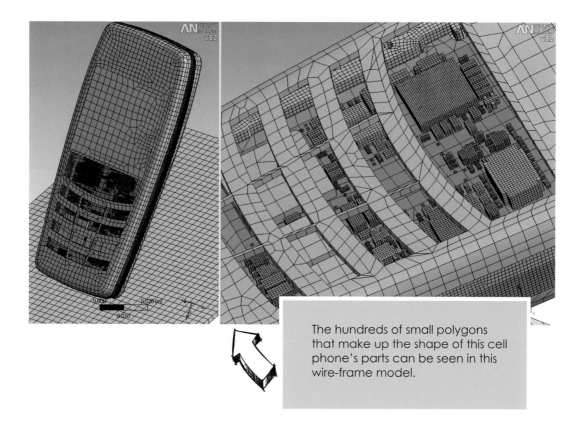

The hundreds of small polygons that make up the shape of this cell phone's parts can be seen in this wire-frame model.

Using CAD, designers can see how parts will look without having to actually make them. Shape, size, color, and texture can be adjusted. These models are eventually passed on for manufacturing real components (see page 30).

OBJECTS AS NUMBERS

In CAD, objects are represented by numbers in a computer's memory. An object is built up from points, called vertices. Numbers show the positions of the vertices. The surface of the object is built up by triangles and other polygons, and the edges of these are made by joining vertices with lines. For example, a cube would have eight vertices (one at each corner). It has six square sides, each with four vertices. Numbers are also used to represent the color and texture of shapes.

Virtual construction

CAD allows designers to build a model of the whole gadget, composed of models of each of the components. The software checks that different parts fit together, as if they were real objects. The parts can then be adjusted, if necessary. Building a computer model of the whole gadget means that many design problems are solved before a real prototype is made (see page 24).

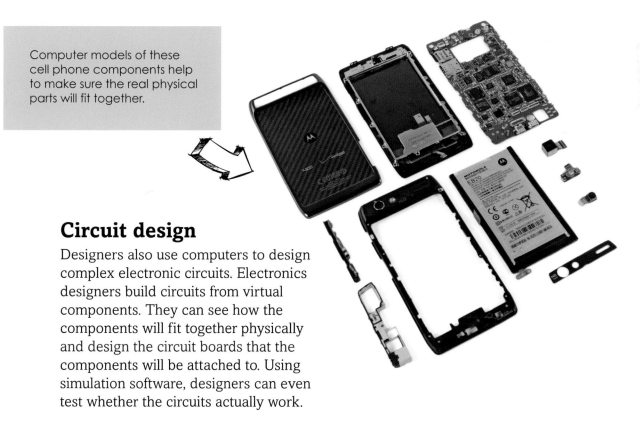

Computer models of these cell phone components help to make sure the real physical parts will fit together.

Circuit design

Designers also use computers to design complex electronic circuits. Electronics designers build circuits from virtual components. They can see how the components will fit together physically and design the circuit boards that the components will be attached to. Using simulation software, designers can even test whether the circuits actually work.

Software design

Just as the physical parts of a gadget need to be designed, so does its software. Software design includes designing the operating system (including screen graphics such as icons and the user interface) and the applications, such as e-mail and mapping. Software design is part of software development, or the software development life cycle.

The first stage in the life cycle, which comes before the designers get to work, is to list requirements and constraints. These include the processing power and memory capacity of the gadget that the software will run on.

Software designers must think about the age and experience of the user. This will influence how they design it.

The design team

Dozens of designers are needed to design a complete operating system, so the work is divided up. Some designers will work on the user interface, others on the sound, and others on the communications (such as connecting to the Internet). The designers organize tasks into modules and write down what each module has to deliver. They do not actually write the software.

Software designers take into account who is going to use the software. For example, an application for young children will require a simple interface. Once the software is completely designed, it can be written, or implemented (see page 26). Games for personal computers, video game consoles, tablets, and smartphones go through a similar software development life cycle (see pages 22–23 for more on game design).

The software development process

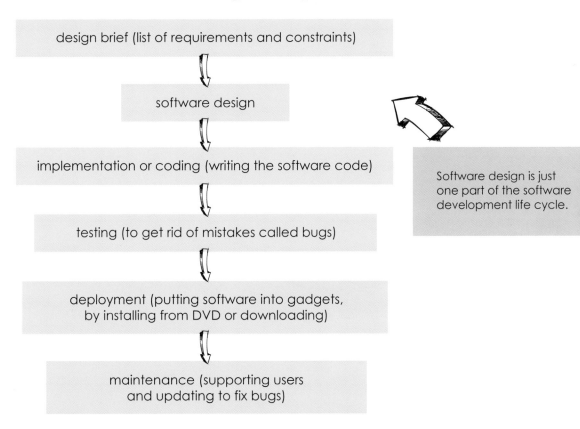

design brief (list of requirements and constraints)

software design

implementation or coding (writing the software code)

testing (to get rid of mistakes called bugs)

deployment (putting software into gadgets, by installing from DVD or downloading)

maintenance (supporting users and updating to fix bugs)

Software design is just one part of the software development life cycle.

CONTRAST THE PAST

Software on the first personal computers and video game consoles was very simple compared to today. This was mainly because computers and consoles had much slower and simpler processors and much less memory than modern computers and consoles. One good comparison is the size of applications. The applications for the Commodore 64, a home computer of the mid-1980s, could be a maximum of 64 kilobytes in size. Today, even a simple application for a smartphone can be 100 megabytes or more in size. That is more than 1,000 times as large.

Game design

Producing games for personal computers, smartphones, tablets, and video game consoles is called game development. The first stage in game development is game design. Once a game developer has decided what genre of game to develop (adventure, simulation, strategy, puzzle, and so on) and decided on the subject of the game, the game can be designed.

Game design depends on the platform (the gadget, console, or computer) the game will be played on. Some games are released for several platforms, and so a slightly different design is needed for each one.

This game designer is using a graphics tablet to manipulate onscreen objects. He has an artist's sketchbook for reference.

People who work on game design

- game designers (figure out storyline, rules, and levels)
- game art designers
- scriptwriters
- movie directors and video editors
- **GAME DESIGN**
- composers and experts in music and sound production
- physics designers (figure out how objects move and interact)

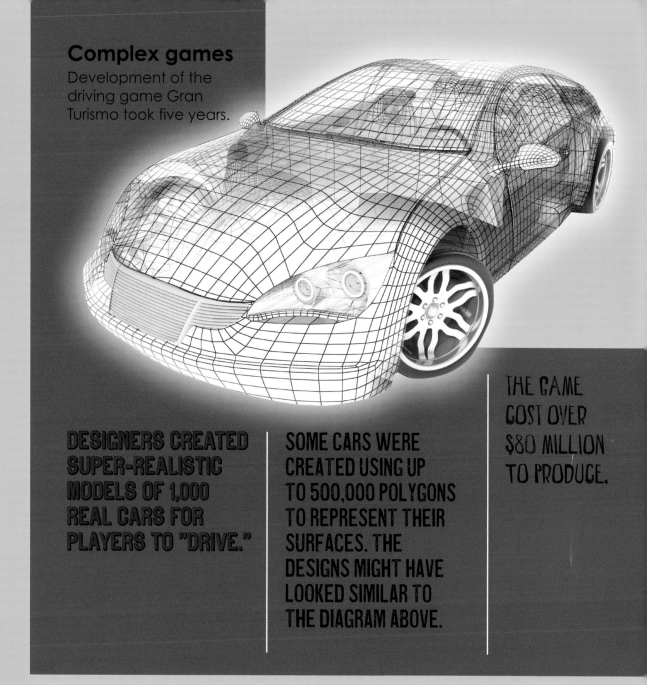

Complex games

Development of the driving game Gran Turismo took five years.

DESIGNERS CREATED SUPER-REALISTIC MODELS OF 1,000 REAL CARS FOR PLAYERS TO "DRIVE."

SOME CARS WERE CREATED USING UP TO 500,000 POLYGONS TO REPRESENT THEIR SURFACES. THE DESIGNS MIGHT HAVE LOOKED SIMILAR TO THE DIAGRAM ABOVE.

THE GAME COST OVER $80 MILLION TO PRODUCE.

WHAT HAVE WE LEARNED?

- A design brief is a list of requirements and constraints.
- Designers choose materials and components.
- A system has inputs, processes, and outputs.
- ICT is used widely in product design.
- Software development includes software design.
- Designers with many different skills are needed for complex products.

PROTOTYPES AND TESTING

Once a product has been designed, engineers build a test version of the product, called a prototype. A prototype shows any problems a designer might have overlooked. Designers make all the parts of the product, such as plastic cases, and source all the standard components, such as batteries and electronic components. They test all the parts individually, then they put the gadget together to test the complete gadget.

Prototype parts can be produced on a 3-D printer, such as this Makerbot Thing-O-Matic. 3-D printing involves making solid objects from a digital file and from layers of material.

Prototype testing

All the systems of the prototype are tested. For example, on a tablet, battery life and communications such as Wi-Fi are tested. Gadgets are also put through a series of physical tests. A cell phone might be dropped onto a solid surface hundreds of times to make sure it does not break. A power supply connector may be plugged into the gadget and removed again thousands of times to test the strength of the plug and socket.

Testing a gadget's software is a complex job, because there may be hundreds of different choices that a user could make when using the gadget, and all of these must be checked.

INTENSIVE TESTING

The JCB company makes tough phones for construction workers. Here are just some of the rigorous tests the phones go through:

- EVERY KEY ON THE KEYBOARD IS PRESSED 300,000 TIMES.
- EACH SIDE BUTTON ON THE CASE IS PRESSED 100,000 TIMES.
- A SIM CARD IS INSERTED AND REMOVED 3,000 TIMES.
- AN SD CARD IS INSERTED AND REMOVED 5,000 TIMES.
- CABLES ARE INSERTED AND PULLED OUT OF THE USB AND EARPHONE CONNECTORS 5,000 TIMES.

Prototype evaluation

A prototype gadget is checked to see if it meets the original requirements. If it does not, some parts or software may need to be redesigned and a new prototype needs to be built. Prototypes may also be given to groups of consumers to test and comment on. This process may be repeated many times before the prototype is finalized. Then, the gadget is ready to go into production.

THINK ABOUT IT

Can you estimate how many times you would press one of the buttons on a video game controller in a year? How often would you press it during a game? Multiply this number by the number of games you play each day, then multiply the answer by 365 to get your answer.

Software coding

System software for gadgets and video game consoles, game software for gadgets and consoles, and other application software must be written and tested before it is released on to the market. Writing software is called coding (and also sometimes software implementation or programming), and it is the job of computer programmers.

Programmers write software in code that a computer understands. The code is a list of instructions for the gadget's processor to carry out. It can be written in one of several languages. Examples of these are C and Java. The language used depends on the gadget the software is going to be used on.

This is what software code looks like. Can you spot "if/then" statements? These help the computer to make decisions when the program is running.

Programmers write code that does the jobs software designers have decided are needed. A manufacturer might employ dozens or hundreds of programmers to write code, depending on the complexity of the gadget or game. Each programmer writes the code for one module of the software. Computers are sometimes used to write some code automatically, and some code is borrowed from software written for other systems.

Some programmers specialize in programming graphics or sound. Graphic programmers have to be expert mathematicians in order to produce realistic and fast-moving 3-D graphics.

LINES OF CODE

To create a big video console game, with 3-D graphics and dozens of levels and characters, programmers will have to write hundreds of thousands—sometimes millions—of lines of code. Operating systems are just as big. The **Android** operating system for smartphones and tablets is made up of more than 12 million lines of code in different languages.

Software testing

Once software is written, it has to be tested. The software must be reliable and robust (which means it must not get off track and make the product stop working). Modules of code are tested individually before being combined to make the finished software. Any problems that are found during testing are called bugs. These can make the software give incorrect results, stop some features from working, or stop a gadget from working altogether.

Reprogramming to get rid of bugs is called debugging. Testing and debugging can take hundreds of hours. Some debugging is done by computer, using programs that automatically spot problems in the software. Bugs are even sometimes put in prototype software deliberately, to check the debugging process itself.

Testing and debugging software

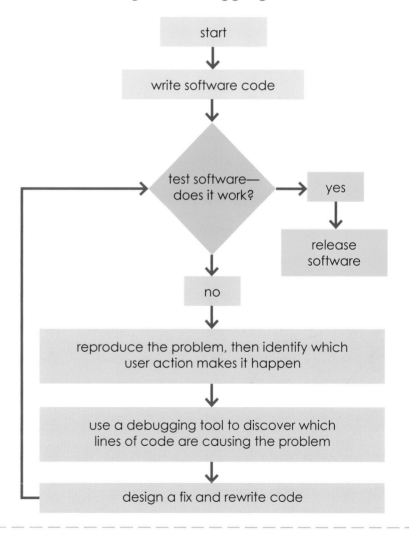

start → write software code → test software—does it work?

yes → release software

no → reproduce the problem, then identify which user action makes it happen → use a debugging tool to discover which lines of code are causing the problem → design a fix and rewrite code

Patents

When manufacturers invent a new technology, they usually apply to the government for a **patent**. A patent office judges whether it really is a new invention. If it is, the office grants the patent to the manufacturer. This stops anybody from copying the invention. It means only a particular manufacturer has the right to make and sell a gadget that contains the new invention.

The idea of patents is to stop one company from stealing another's ideas. Companies can apply for patents for both software and hardware. There are often legal battles over patents. For example, in 2011, Apple and HTC argued over features of their touch-screen interfaces. However, companies often allow others to use their patents for a fee. Patents are also shared when gadgets need to use the same technologies to communicate (such as Bluetooth).

Drawings form part of the patent application for devices that feature new ideas and technology.

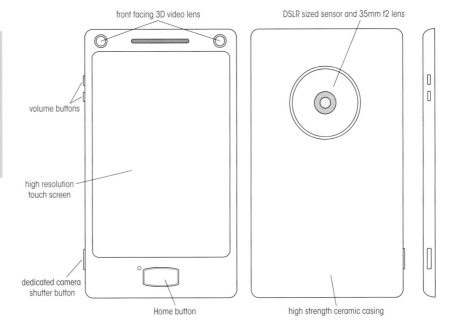

front facing 3D video lens

DSLR sized sensor and 35mm f2 lens

volume buttons

high resolution touch screen

dedicated camera shutter button

Home button

high strength ceramic casing

VIDEO GAME RATINGS

Video games must have an age rating on the packaging, and the rating must be displayed on a web site if the game is downloaded. The age rating shows what age of player the game is appropriate for. It helps stop children from playing games that have content (such as violence) that is not suitable for them. Game designers think about a player's age as they design a new game. In the United States, age ratings are awarded by the Entertainment Software Rating Board (ESRB).

Safety standards

All gadgets that are sold must be safe for people to use. This is very important—gadgets work with electricity and contain materials that could be toxic if they leak out. Manufacturers are responsible for making their products safe, especially since many users will be young children. In the United States and many other countries, electronic device manufacturers need to meet certain safety requirements before they can sell a product. Gadgets must also carry safety warnings where necessary. For example, gadget batteries must have a warning against taking them apart, because they contain hazardous chemicals.

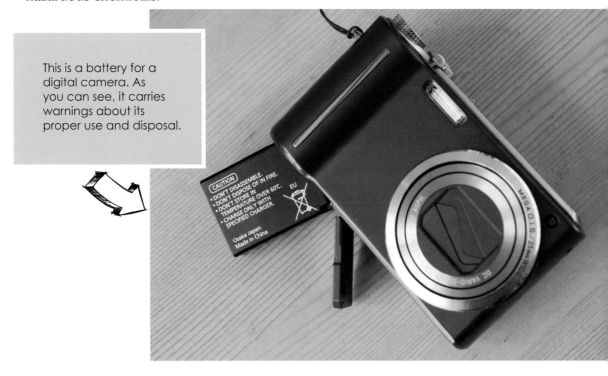

This is a battery for a digital camera. As you can see, it carries warnings about its proper use and disposal.

WHAT HAVE WE LEARNED?
- A prototype is built to test that a product will work properly.
- A product may be redesigned if the prototype does not meet the original design requirements.
- Writing software is known as software coding.
- Software is debugged to remove errors.
- A patent protects new technology from being copied.
- Computer games have to carry an age rating.
- Gadgets must carry safety warnings.

PRODUCTION

Once a design specification is complete, thoughts can turn to production—which is where the gadget turns from an idea into reality. The first stage of production is to write a manufacturing specification for all the different components, from the complex integrated circuits to the screws that hold the case together. Some components, such as the plastic or metal case, will be specially made. Others, such as memory chips, will be existing parts that are sourced from other manufacturers.

Making tools

Before production can begin, tools must be made. These tools include molds for making plastic parts such as buttons. A factory must also be set up. This could be an existing factory or a completely new one that will need to be designed and built. Gadgets are frequently made in a different country from the one where the manufacturer is based, normally because labor costs there are lower. This is known as remote manufacturing.

Computer-aided manufacturing (CAM)

Lathes, milling machines, and drills are used to shape plastics and metals into components. These machine tools are controlled by CAM and automatically

cut and drill away material. They are programmed with computer numerical control (CNC) and cut very accurately. Data from CAD files is fed to the machine tools, which make the components automatically. Designers think about how components will be manufactured as they design them, so that the components can be made easily and with as little waste as possible.

This is a numerically controlled cutting table. It is cutting components out of a sheet of steel.

Sourcing materials

The materials used to make the components for gadgets have to be sourced from somewhere. For example, aluminum, steel, gold, and silver are all used in gadgets. They are found in materials called ores, which are dug out of the ground in mines. The ores are processed to extract the metal—in the case of steel, iron is extracted and mixed with another element, usually carbon. The metal is then shaped into rods, bars, and sheets, ready for manufacturing. Most plastics are made from oil, which is also extracted from rock. Some gadget manufacturers make use of recycled plastics and metals, to reduce environmental damage.

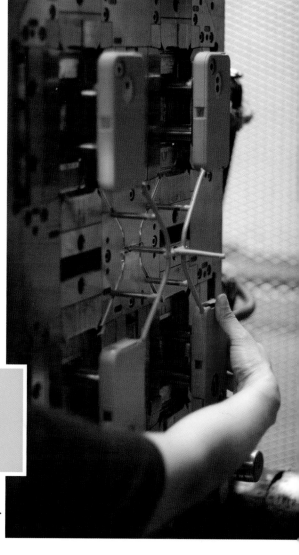

This is one part of a molding machine, from which brand new iPhone cases are being removed. The central section is where plastic is injected into the mold.

ECO IMPACT

Metals known as rare-earth metals are needed to manufacture many components of gadgets, including touch screens. They include tantalum and yttrium. The ores of these metals are hard to find. In 2011, 97 percent were mined in China. Toxic chemicals, including strong acids, are needed to process the ores, and in some places these have leaked out of processing plants, harming the environment. Some ores also contain radioactive materials, which are hard to dispose of.

Making integrated circuits

All gadgets need electronic circuits to make them work. Most are contained in integrated circuits, also known as silicon chips. An integrated circuit is made of silicon with microscopic electronic components built on its surface. Integrated circuits normally do a specific job. In a gadget or video game console, there will be one or more microprocessor chips, some memory chips, chips that produce graphics, and chips that produce sound.

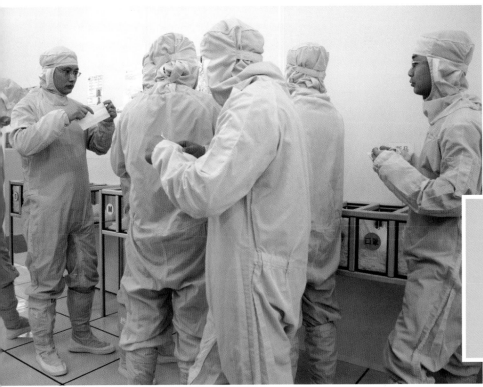

People who work in chip-manufacturing areas of gadget factories must wear special suits to keep the work space free of dust and dirt.

Dozens of integrated circuits are built in a grid pattern on a wafer of silicon up to 12 inches (30 centimeters) across and 1 millimeter thick. Silicon is used because it is a semiconductor—a material that has properties for conducting electricity between those of an insulator and those of a conductor. How well silicon conducts or insulates can be altered by adding other substances.

The components and connections are built up in layers on the wafer by a series of processes. Some processes add materials, such as metals, insulators, and semiconductors. Some change the electrical properties of the materials, and others remove materials from certain parts of the wafer. Specks of dust, high temperatures, and high humidity can ruin integrated circuits, so they are manufactured in a dust-free area called the factory's clean room. Here, temperature and humidity are carefully controlled.

Manufacturing an integrated circuit

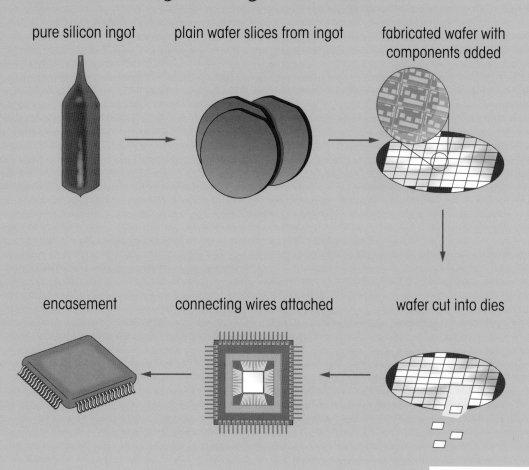

pure silicon ingot

plain wafer slices from ingot

fabricated wafer with components added

encasement

connecting wires attached

wafer cut into dies

This flow diagram shows the main steps involved in manufacturing a finished circuit from silicon.

HOW SMALL?

The tiny components on integrated circuits are measured in nanometers. How big is a nanometer?

- A millimeter equals a thousandth of a meter. (A meter is about 3.3 feet.)
- A micrometer equals a thousandth of a millimeter (or a millionth of a meter).
- A nanometer equals a thousandth of a micrometer (or a billionth of a meter).

A human hair is about 100,000 nanometers wide!

Mass production

All gadgets are made using mass production. In this process, thousands, sometimes millions, of the same product are made together. Each individual product (called a unit) is exactly the same. Mass production makes each unit cheaper than if just a few were built. The cost saving comes because materials and components are bought in bulk, which makes them cheaper.

Mass production takes place on an assembly line. This means parts are added to a product as it moves through the factory. At the end of the line, it is complete. This is a very efficient way of making products. The more products made on an assembly line, the cheaper it becomes to make each unit. This falling cost is known as economy of scale.

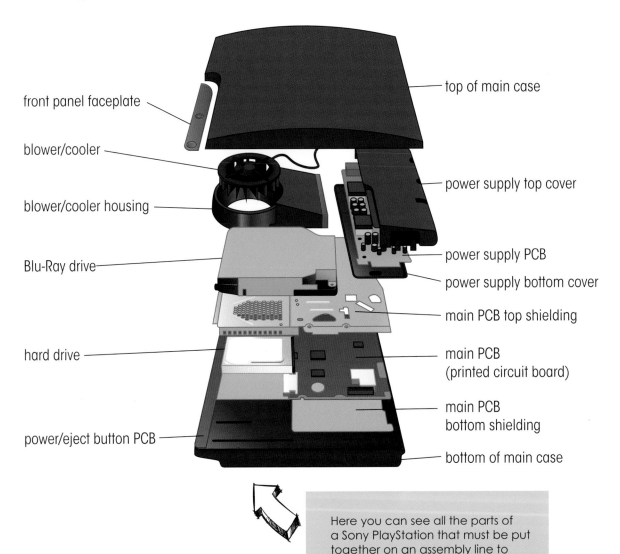

front panel faceplate

blower/cooler

blower/cooler housing

Blu-Ray drive

hard drive

power/eject button PCB

top of main case

power supply top cover

power supply PCB

power supply bottom cover

main PCB top shielding

main PCB
(printed circuit board)

main PCB
bottom shielding

bottom of main case

Here you can see all the parts of a Sony PlayStation that must be put together on an assembly line to create the finished product.

As a gadget moves along an assembly line, parts are added by workers and by robots. Each worker has a role on the assembly line, and many do skilled jobs, such as soldering electronic parts together. Robots are used to position parts accurately and quickly and may also do jobs such as soldering instead of workers. Workers make sure the assembly line is supplied with components. Some assembly lines run continuously, producing gadgets 24 hours a day.

QUALITY ASSURANCE AND CONTROL

Manufacturers operate a quality assurance system to ensure that the gadgets people buy are of high-quality. When each item is complete, it is examined and tested by a quality control team. Automatic machines that test all the different features of the gadget at high speed may be used.

Assembly line workers must repeat the same task over and over again.

CONTRAST THE PAST

Mass production was developed hundreds of years ago, but automated assembly lines with robots were developed in the 1970s. Before then, all jobs on assembly lines were carried out by people, who could get tired and make mistakes. Robots and other automatic machines repeat tasks with the same accuracy and speed on every gadget that passes down the line. They have transformed manufacturing and improved the quality of products. However, people are still needed to make and maintain the robots and to carry out tasks that robots cannot do.

Software production

When software has been coded, it is just a computer file or collection of files, not a physical product. System software is loaded onto gadgets after they are produced. It is either loaded into a gadget's memory or onto its hard drive. Software that customers buy is either supplied on DVD or Blu-Ray discs, or it is downloaded from a manufacturer's web site.

Increasingly, software is downloaded rather than supplied on disc. Unlike discs, which need to be manufactured, no production is required for downloaded software. The computer files are simply copied from one computer to another.

On DVD and Blu-Ray discs, data is represented by microscopic pits in the disc surface. This is how a DVD is made:

- Molten plastic is forced into a mold containing a metal master disc.
- A plastic disc comes out of the mold with a pattern of pits on it.
- The plastic disc is coated with a thin layer of aluminum.
- A layer of protective plastic is added to the aluminum.
- A title is printed on the reverse side of the disc.

This is a DVD pressing machine in action. The process is completely automatic, and each DVD is exactly the same.

Product packaging

Packaging protects a gadget while it is transported from a factory to a store or to a customer, and while it is being stored. The packaging for a product also contains items such as a power supply, cables, and printed booklets. Packaging has to be designed and manufactured in time for gadgets to be packaged as they come off an assembly line. Packaging is made from paper, cardboard, and plastics.

BATCH PRODUCTION

DVDs and Blu-Ray discs are made by batch production. This means production is not continuous, as it is in mass production. A certain number of items are produced, then production stops and a different item is produced. For example, 10,000 DVDs of a game might be produced first, then 10,000 more when stores run out of stock.

Designers should try to use as little packaging as possible and use recyclable or biodegradable materials.

WHAT HAVE WE LEARNED?

- A manufacturing specification describes the parts of a gadget, what materials they are made from, and how they will be manufactured.
- In CAM, machine tools are controlled by computer, allowing them to make components very accurately.
- The metals and plastics used to make gadgets come from rocks or are made from oil.
- Gadgets are made by mass production on assembly lines.
- A quality assurance system ensures gadgets work properly when they leave the factory.

MARKETING

Marketing is about promoting and selling products. All gadget manufacturers have marketing departments that are responsible for making sure consumers know when new products are released and what advantages they offer.

Advertising

The main aim of advertising is to persuade people to part with their money and buy a new product. It is very important for manufacturers to sell as many units as possible, so they can make a profit and continue to produce goods. They often employ advertising agencies to run campaigns for them. Manufacturers or agencies will plan where and when to advertise. They will produce material for television, web sites, newspapers, and magazines. Advertisements are targeted at particular sections of the population, such as young people or wealthy professionals. It costs a lot of money to run a big advertising campaign, so a budget will have been set aside.

Advertisements are designed to show off the most desirable features of a product. These include how user-friendly it is, how much fun it is, its long battery life, the size of its screen, its speed, and its software, such as apps that allow access to social networks.

Despite the rise of internet advertising, some manufacturers still believe printed ads are very effective. Here, a boy walks past a huge indoor advert in Seoul, Korea.

CONTRAST THE PAST

In the past, gadget manufacturers advertised in the print media, on television, and on billboards. In the 1990s, people began to connect to the Internet at home. This was the start of a revolution in marketing. Today, manufacturers advertise through their own web sites, through search engines, and with banner ads on related web sites. They also send e-mails to potential customers and maintain a presence on social-networking sites such as Facebook and Twitter.

SUCCESSFUL MARKETING

Amazon has sold millions of its Kindle e-book reader. Its clever advertising campaign emphasized the features that make the Kindle different from tablets, such as its low cost, e-ink screen, and lightness. Amazon also advertised the device on its own web site, which gets millions of hits a day.

You can scan an ad's quick response (QR) code with a smartphone. The code will connect the smartphone's browser to the product's web site.

Planning a campaign

Marketing begins with market research (see page 10). This takes place before a gadget is actually being manufactured. Advertisements need to be made and ready to appear just before the product becomes available to buy, so that consumers are aware of it and possibly looking out for it. Because most gadgets have a short life cycle, it is important there is no delay between a gadget first coming off the assembly line and sales being made. Otherwise, a rival manufacturer might capture the interest of the consumer and make a sale instead.

Product launch

When a gadget or game is ready to sell, its manufacturer organizes a product launch. This is an effective way to raise awareness of a new product and to create a sense of excitement around it. In the lead-up to the launch, newspaper and television ads will appear and a web site will go live.

Manufacturers may also organize a launch event, where the new gadget or game is shown off to experts and the media. They can try it out and write reviews of it, which also helps to raise consumer interest. A good product launch will create a high demand for the product when it is released.

This enormous crowd is waiting for the launch of the Apple iPhone 4S.

FAIL

Not all gadgets are a success. In 2006 Microsoft launched the first of its Zune range of portable media players to compete with Apple's iPod. Although many reviewers said the Zune was a good product, especially some of the software, it did not have enough marketing support to compete with the established iPod. The Zune players were withdrawn in 2011.

Gadget and video game fans eagerly wait for the launch of a new product. It is important that there are enough units of a gadget or game available when it is launched, because otherwise demand may outstrip supply. This means supplies will run out and consumers will have to wait before they can buy—with the possibility they will turn to a competitor's product instead.

Sales figures measure how many units are bought each week, month, or year. When first released, the figures begin to rise. They will usually rise for a few months, then they stop rising during the maturity phase. They will then begin to fall until the product is no longer available, when they hit zero. This rise and fall in sales is part of a product's life cycle.

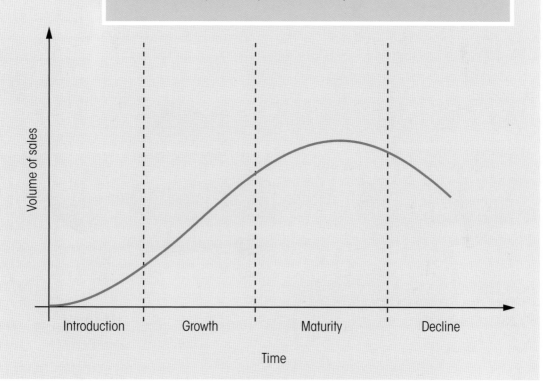

WHAT HAVE WE LEARNED?
- Marketing is about advertising and selling products.
- Successful marketing encourages consumer interest.
- A product launch can create high demand.

PRODUCT MAINTENANCE

A company's involvement in a product's life cycle does not end on the day the product is sold. Product maintenance comes next. This means keeping a product working while it is being used—this stage is called its useful life. Like all other products, gadgets should be made to a good standard. They should be reliable, which means they should work for a reasonable length of time without malfunctioning or breaking. By law, manufacturers have to guarantee their gadgets for a certain amount of time (normally a year). This is written in a product's warranty.

Manufacturers also provide support for their customers, to help those who are having problems making their gadgets work properly. Online support is provided on a manufacturer's web site. There will be a list of common questions, known as frequently asked questions (FAQs). Customers may also be able to talk to a member of a support team.

A repair engineer is using an electronic test meter to check the circuits of a malfunctioning smartphone.

PRODUCT RECALL
Occasionally, a product might not work properly or may develop a problem that makes it unsafe to use. In these cases, the product is recalled. This means customers can send their gadgets back to the manufacturer to be repaired or replaced.

Spares and repairs

Manufacturers make spare parts for their gadgets, which are used to replace parts that may get damaged or broken accidentally. Because gadgets have a short life cycle, their parts are rarely used long enough to wear out. Normally, the only part that might have to be replaced in a gadget is the battery, which can lose its ability to hold a charge after being recharged a few thousand times. If a gadget does develop a problem, it can usually be repaired by the manufacturer.

All batteries are marked with a recycling symbol to remind people to take them to a recycling center like this one after they are used up.

ECO IMPACT

Batteries contain many toxic chemicals, including metals such as lead, cadmium, nickel, and lithium. If batteries are thrown into landfill sites, these chemicals can leak into the soil and into water in the ground, with possible damage to plants and animals. So, when batteries in gadgets are replaced, the old batteries should be taken to a recycling center. Then, some of the materials can be extracted and used again.

Software updates

Gadgets and video game consoles could not work without their software. Software does not break down the way physical parts can, but it can have irritating bugs in it that are not spotted at the software testing stage of the life cycle. These are often spotted by users and reported to the manufacturer. Sometimes different applications can stop each other from working properly.

Reasons for software updates

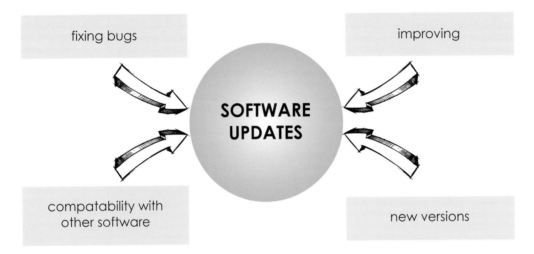

fixing bugs

improving

SOFTWARE
UPDATES

compatability with
other software

new versions

Bugs mean that software sometimes needs to be rewritten slightly to make it work properly. This is called a software update. Sometimes the update is just a small piece of code, called a patch. At other times, the software is completely replaced with a new version. An update is shown in the software version number (for example, version 2.1 becomes version 2.2). New versions of software normally solve several bugs all at once and may add new features. Software updates lengthen the life cycle of a gadget, because they improve the gadget with a simple operation.

AUTOMATIC UPDATES

Software updates often happen automatically. A gadget checks the manufacturer's web site every few days to make sure its software is up-to-date. If an update is available, the gadget downloads it, after checking with the user. Minor software updates are free, as they are needed to keep a gadget working properly. Major updates normally have to be paid for.

Malware

Malware—such as viruses, worms, and trojan horses—is software that damages a computer or stops it from working properly. Viruses and worms spread in e-mails or over computer networks. Trojan horses do not spread, but they get into a computer unnoticed by the user and do damage.

Gadgets such as smartphones and tablets are really computers, and so people are writing malware for them. Malware is a problem for gadget users and costs time and money to get rid of. To avoid malware, always be careful which apps you download and think about installing anti-virus software.

How malware is spread

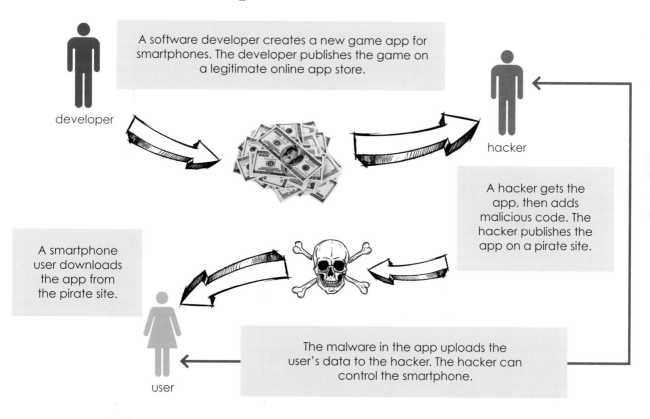

developer

A software developer creates a new game app for smartphones. The developer publishes the game on a legitimate online app store.

hacker

A hacker gets the app, then adds malicious code. The hacker publishes the app on a pirate site.

A smartphone user downloads the app from the pirate site.

The malware in the app uploads the user's data to the hacker. The hacker can control the smartphone.

user

WHAT HAVE WE LEARNED?
- Product maintenance keeps a gadget working after it has been sold.
- Manufacturers produce spare parts for mending their products.
- Software updates fix minor bugs in software.

END OF LIFE

A gadget's life cycle comes to an end when it is no longer used. Consumers stop using millions of gadgets every year. They become electronic waste (e-waste). So, what happens to them?

Some are thrown away, and so come to the end of their lives. Others end up sitting in a drawer until they become obsolete. This is a waste—the life cycles of gadgets can be extended by re-using them. When a gadget does reach the end of its life—either obsolete, worn out, or broken—its components and materials can be recycled.

Reusing gadgets

A gadget can be reused by passing it on to a family member or friend. There are also many charities that refurbish old gadgets. Refurbishment means getting a gadget ready to be sold to a new user. In the case of a cell phone, for example, the charity checks that the phone has not been stolen, repairs it if necessary, cleans and repackages it, and then sells it.

Reusing games

Video games have a different end to their cycle. Downloaded games stay on computers or devices until they are deleted. Games on DVDs or Blu-Ray discs are normally passed on or sold to other users, sometimes many times. They have life cycles that last many years.

Old cell phones can be sent to countries where there are few landlines and where many people cannot afford the latest gadgets.

REFURBISHMENT

In addition to charities, some gadget manufacturers will refurbish old cell phones, too. Some cell phones are known as "beyond economic repair," meaning they cannot usefully be reused. However, about 95 percent can be refurbished. If you do pass on a cell phone in this way, remember to delete as much personal data as possible before you do.

ECO IMPACT

If a gadget is simply thrown away at the end of its life, the energy and materials used to make it are lost, which is bad for the environment. Here are some eco-friendly options that manufacturers and users can choose:

End of life

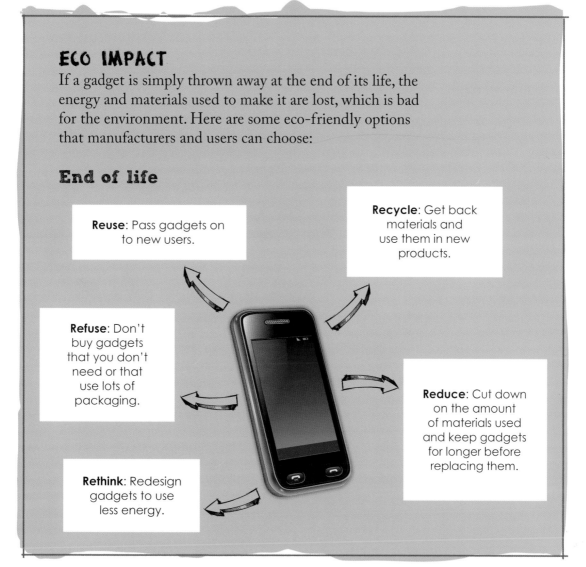

Reuse: Pass gadgets on to new users.

Recycle: Get back materials and use them in new products.

Refuse: Don't buy gadgets that you don't need or that use lots of packaging.

Reduce: Cut down on the amount of materials used and keep gadgets for longer before replacing them.

Rethink: Redesign gadgets to use less energy.

Recycling

When a gadget is recycled, it is broken down into its components. The materials in the components are extracted and become raw materials for new products. Gadget design should make it easy for the device to be broken up at the end of its life. In most cases, about 80 percent of the materials in a gadget can be recycled. The phone manufacturer Nokia claims that 100 percent of the materials in its phones are recyclable.

The materials we can recover from gadgets are mostly plastics and metals. Touch screens and integrated circuits use metals that are rare and hard to find in rocks. This means recycling them is very important.

These self-assembling cell phones are at the end of their lives. They are being heated to break down their plastic parts, which are then ready to be recycled.

Recycling methods

You cannot sort the parts of a gadget into different materials yourself, as you do with the recyclable materials in household waste. You should take gadgets to your local recycling center. Specialized recycling companies carry out the recycling process.

We have finally reached the end of the life cycle of a gadget. However, the materials used to make it may live on in new products. Its components, including electrical connectors, screens, keyboards, camera lenses, and speakers, may be used again, too. Cardboard and paper from gadget packaging can be recycled. These materials and components enter a new life cycle.

LANDFILL

Unfortunately, many people are unaware that gadgets can and should be recycled. Millions of old and broken gadgets still end up in landfill, wasting the energy and materials used to make them. The gadgets will take thousands of years to break down and will then leave potentially hazardous chemicals in the ground. This is the worst thing that can happen to a gadget at the end of its life.

This is the WEEE (waste electrical and electronic equipment) Man— a 23-foot (7-meter), 3.3-ton sculpture at the Eden Project in Cornwall, England. It represents the amount of this kind of waste the average British household throws away in a lifetime.

WHAT HAVE WE LEARNED?
- End of life is when a gadget is obsolete or broken beyond repair.
- Reusing or refurbishing a gadget extends its life cycle.
- Most of the materials in a gadget can be recycled.
- Many gadgets still end up in landfill sites.

TIMELINE

1913   The Ford Motor Company perfects a moving assembly line for the production of the Ford Model T car.

1958   The integrated circuit is invented.

1958   Hard-wearing plastic called polycarbonate is invented.

1961   The first industrial robot, a UNIMATE, starts work in the General Motors car factory in Ewing Township, New Jersey.

1963   The first CAD system is developed.

1970s   Light-emitting diodes (LEDs) are first used in gadgets for displays and indication lights.

1972   The first electronic game console, the Magnavox Odyssey, is released. It plays simple bat-and-ball games on a television screen.

1974   The first transparent touch screen is developed.

1978   The first cell phone network opens in the United States.

1989   The Nintendo Gameboy, the first handheld video game console, is released.

1990s   The second generation of mobile networks, known as **2G**, opens, allowing for the development of small cell phones and the introduction of text messaging. Also, people begin to connect to the Internet from home.

1991   The first Wi-Fi networks are developed for use in stores.

1993   The Apple company releases the Newton, a personal digital assistant (PDA) with a touch screen.

1994   The Entertainment Software Rating Board (ESRB) is formed to ensure that the content of video games is appropriate for different audiences.

1994   The first generation Sony PlayStation video game console is released.

1996   The digital versatile disc (DVD) is introduced.

1998   The first MP3 music players are released.

1998   The driving game Gran Turismo for the PlayStation is released.

2000s   The third generation (**3G**) mobile networks open, allowing mobile devices to download data.

2001   The first generation Microsoft Xbox is launched.

2001   Apple begins to sell its first MP3 player, the iPod.

2002   The Blackberry is one of the first smartphones available to buy.

2004   The first Nintendo DS handheld console is released, as is the PlayStation portable (PSP).

2005   The Xbox 360 is released.

2006   The Nintendo Wii console is released, with its revolutionary motion control system.

2006   The PlayStation 3 is released.

2007   The first Apple iPhone is released.

2007   The Kindle is released by Amazon.

2007   The Android operating system is released for mobile devices.

2010   The Xbox Kinect is launched, allowing players to control games with body movement.

2010   The first Apple iPad is released.

2010   The Nintendo 3DS is the first portable console with a 3-D screen.

GLOSSARY

2G, **3G**, and **4G** generations of mobile data networks, which have allowed mobile devices to have increasing speeds of communication and features (such as text messaging, media messaging, and web browsing)

Android operating system used on many smartphones and tablets

application (app) software that makes a gadget do a specific job

assembly line system of mass production in which products are assembled from their components as they move along a conveyor belt

cloud computing system in which data and applications are delivered to devices via the Internet, rather than being stored on the device itself

component part of something bigger

computer-aided design (CAD) use of computer technology for the process of designing objects

computer-aided manufacturing (CAM) use of information and computer technology in the manufacturing of products, normally in conjunction with CAD

constraint limitation on how something is made

design process of choosing the materials, components, and appearance of a gadget

engineering use of mathematics and science to design and build structures, machines, devices, systems, and processes

input information that goes into a system

integrated circuit electronic circuit made up of microscopic components built into a slice of silicon

manufacturing process of actually making products from materials and components

marketing bringing a gadget or game to the attention of the consumer; to try to sell the gadget

mass production process of manufacturing identical products in large numbers

operating system software that controls the parts of a gadget itself. Examples include Apple's iOS and Google's Android.

output information that comes out of a system

patent authority from a government that allows a manufacturer sole right to make a product or component

process what happens to information between the input and output of a system

product life cycle series of events in the life of a gadget, from initial idea through to design, manufacturing, useful life, and disposal or recycling

prototype one of the first examples of a product to be made, so that it can be tested to make sure it works properly

recycling processing old materials and products to make new materials and products. Recycling saves raw materials and reduces waste.

requirements in design, necessities affecting the cost or function of a final product or service

smartphone cell phone, normally with a color touch screen, that allows voice calls, texting, e-mailing, and web browsing and runs other applications

software instructions and data that control what a gadget or video game console does

software design figuring out how software for gadgets and games will work

specification list of requirements and constraints for a product

system set of things that work together to do a job. All systems have an input, a process, and an output.

technology modifying objects and materials to satisfy people's needs and wants

useful life part of the life cycle of a product when the product is being used

FIND OUT MORE

Books

Funk, Joe. *Hot Jobs in Video Games: Cool Careers in Interactive Entertainment*. New York: Scholastic, 2010.

Gifford, Clive. *Cool Tech: Gadgets, Games, Robots, and the Digital World*. New York: Dorling Kindersley, 2011.

Morris, Neil. *Gadgets and Inventions* (From Fail to Win: Learning from Bad Ideas). Chicago: Raintree, 2011.

Web sites

www.bls.gov/k12/computers.htm
careerplanning.about.com/od/occupations/a/videogamecareer.htm
www.discoverengineering.org
Look at these web sites for information on how to start a career in the world of engineering, technology, and gadgets and video games. It is useful to think ahead to make sure you end up doing something that interests you!

greenergadgets.org
The web site of Greener Gadgets is full of information about how to be more "green" with your technology. Among its many features is a locator that allows you to type in your zip code and find where you can recycle gadgets. The site is also full of information about recent developments in recycling.

www.nobelprize.org/educational/physics/integrated_circuit/history
Learn about the history of integrated circuits from the official web site of the Nobel Prize. Jack Kilby, who invented the integrated circuit, won the Nobel Prize for Physics in 2000.

www.nokia.com/global/about-nokia/people-and-planet/
sustainable-devices/recycling/recycling
Discover more about the importance of recycling your cell phone from Nokia. This page includes a guide on how to recycle your old phone.

Places to visit

Museum of Science and Industry
57th Street and Lake Shore Drive
Chicago, Illinois 60637
www.msichicago.org
The Museum of Science and Industry has fascinating exhibits about
technology, including the Networld exhibition, which explores the Internet.

The Tech Museum
201 South Market Street
San Jose, California 95113
www.thetech.org
The Tech Museum is filled with interactive exhibits exploring all kinds of
technology, including gadgets.

Topics for further research

- Gadget history: The development of electronic gadgets and games is a
  fascinating topic to research. Think of the factors that have influenced
  the features of gadgets (for example, many gadgets now make use of
  the Internet).

- The future: What do you think the world of gadgets might be like in
  50 years? Remember, this is a very fast-moving field. Think about what
  features you would like to have in a gadget of the future.

- Careers in gadgets: Think about which part of the gadget life cycle
  you would like to be involved in (such as electronic design or
  manufacturing). Useful subjects to study in school include math, science,
  and computer technology. Look at page 54 for some helpful web sites.

- Sustainability: Find out more about an aspect of gadget sustainability, such
  as self-dissembling devices and the technologies involved. Do you and your
  friends think about the environmental impact of buying a new gadget and
  throwing away an old one? If so, would it influence your decision to buy
  the new smartphone you really want?

- Latest trends: At the time you are reading this, what are the latest trends
  in gadget and video game technology? What is the latest "must-have"
  smartphone feature, game technology, or application?

INDEX